Who has mo
It's easy to c(

Are there more fish here or more fish there?

Which building has fewer windows?
Which building has many more?

Are there more elephants in the water or standing on the shore?

Are there more puppies or more kittens?

Or is the number just the same?

Which sign has more letters?
Which has the shortest name?

Which tree has more leaves?

Which tree has less?

Are there more than twenty candies?

Go ahead and take a guess!

Now you know how to compare—
use more or less anywhere!